Photosynthesis

Advanced Biology Study Notes for Teachers & Students

Andrew Williams, Ph.D.

Version 1.2

Release Date: 24th October 2013

Table of Contents

Use of This Book & Accompanying Files

This book includes a mock exam at the back. You can also download the exam from my website, and the images used in this book (see the end of the book for the download link).

The downloadable examination comes as a PDF for easy printing, but also as a Word document, which you may edit for your own purposes within the classroom. All images included in the download may also be used within the classroom, e.g. on worksheets.

The examination, mark scheme and images may not be sold or used in any other material, online or in physical print.

About These Photosynthesis Study Notes

Photosynthesis is a topic that many new teachers and students struggle with at advanced levels. These notes are designed to make your life a little easier, by highlighting the main points in photosynthesis that examining boards expect advanced level students (16-18 years old) to know.

It's impossible to create study notes that will match every curriculum out there, so I haven't tried. What I have endeavoured to do is write a book that covers the core material of the subject, and will therefore provide a structure to teach or learn this challenging topic. Ultimately, if you understand the concepts in these notes, you will understand photosynthesis to a good level.

The main ideas you should think about when learning photosynthesis at this level are as follows:

1. The need for energy in every living thing.
2. Where does that energy ultimately come from?
3. The structure, synthesis, and role of ATP.

These notes will cover a lot of ground in as little space as possible, using diagrams and summary points. All of the main diagrams are available as downloadable jpeg files – see section 6 of this book.

The book on photosynthesis has three main sections.

1. Photosynthesis theory
2. Practical work
3. Practice examination (with mark scheme).

Here is a summary of what's covered in these sections:

Photosynthesis Theory

- The structure of a dicotyledonous leaf, palisade cells, and a chloroplast - relating structure to function.

- The photoactivation of chlorophyll resulting in the conversion of light energy into ATP and reduced NADP.

- The processes of cyclic and non-cyclic photophosphorylation.

- The Calvin cycle involving the light-independent fixation of carbon dioxide, combining it with a 5C compound (RuBP) to yield two molecules of GP (PGA), a

3C compound, and the subsequent conversion of GP into carbohydrates, amino acids and lipids.

- How the light-dependent reaction, and the light-independent reaction are linked.
- Limiting factors in photosynthesis, which is a nice introduction to the practical work.

Practical Work

- Investigations on the effects of limiting factors, such as light intensity, CO_2 concentration, and temperature on the rate of photosynthesis; compensation point; gas analysis.

End of Topic Exam

- A complete end of topic exam and teacher mark scheme.

1. Introduction - Energy

Energy is the ability to do work.

1.1 Animals hunt their food, plants cannot – they have to make it.

It is important for students to appreciate that all of the energy found in living organisms ultimately originated from the sun. Students understand this for green plants - they capture the sun's energy and produce glucose as **potential chemical energy** during photosynthesis. However, students often fail to appreciate that all of the energy in any animal also comes from the sun.

Animals eat plants to get energy and those animals that don't eat plants, eat animals that have eaten plants.

For the sun's energy to enter the living world, evolution had to come up with something quite special. Plants, driven by their inability to move and hunt for food, evolved mechanisms to make their own food from freely available simpler compounds – carbon dioxide and water.

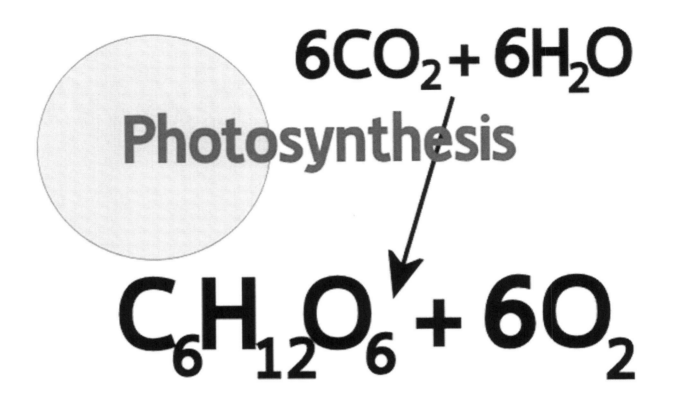

This process requires the energy from the sun.

1.2. Metabolic reactions

There are thousands of chemical reactions taking place in an organism, mostly catalysed by specific enzymes. These life-sustaining reactions are collectively known as **METABOLISM**.

Metabolism can be divided into two groups of reactions – those that build up substances (**anabolic reactions** –students may know about anabolic steroids in bodybuilding), and those that break down substances (**catabolic reactions**).

Anabolic Reactions	Catabolic Reactions
Complex compounds built from simpler ones.	Breakdown of complex compounds into simpler ones.
Requires energy	Releases energy
Examples: from the synthesis of simple molecules like sugars (in photosynthesis), amino acids and fats, to macromolecules like proteins, fats, starch, cellulose and glycogen.	Examples: breakdown of sugars and fats in respiration.

In addition to anabolic reactions, energy is also required for life processes: movement, growth, repair, active transport, maintaining a constant internal environment (homeostasis, e.g. keeping warm), and so on and so forth.

1.3 Adenosine Triphosphate (ATP)

ATP is the currency of energy used by living cells. The structure should be familiar:

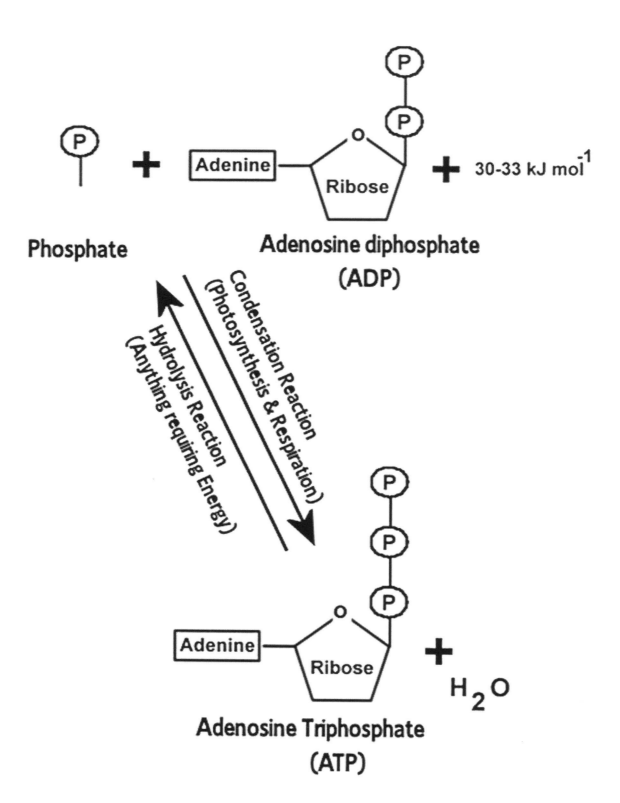

ATP is made from the same building blocks as our nucleic acids (DNA & RNA) – a pentose sugar, inorganic phosphate, and a nitrogenous base.

Occasionally ADP can also be hydrolysed to AMP (adenosine monophosphate), releasing a similar amount of energy. AMP can be further hydrolysed to adenosine, and phosphate, releasing small amounts of energy, but this is insignificant in metabolism.

High Energy Bonds

ATP is often described as a high energy molecule, or as containing high energy bonds, that when broken release energy. However, the covalent bonds in this molecule are not special, and it is incorrect to describe the bonds as high energy. Think of ATP as a dehydrating agent – ATP pulls water from other reactions and is hydrolysed in the process.

2. Plant Adaptations to Photosynthesis

Photosynthesis is a complex biochemical process in which molecules of carbon dioxide and water are combined using the energy from the sun to produce glucose. We will see that the whole process can be divided into two groups of reactions – those requiring light (light-dependent reactions) and those that do not require light (light-independent reactions).

For now, a simple overall equation for photosynthesis will suffice:

$$6CO_2 + 6H_2O \xrightarrow[\text{chlorophyll}]{\text{Sunlight}} C_6H_{12}O_6 + 6O_2$$

NOTE: Make sure your students are happy balancing equations and can see why 6 molecules of carbon dioxide, and 6 molecules of water, are required to make one molecule of glucose (with 6 molecules of oxygen given off).

From this equation you can remind your students about limiting factors. They should be able to spot four limiting factors immediately, and with a little thought, a 5th one too.

This equation does not tell us the true story of the complex reactions taking place in the production of glucose, but it is a useful starting point as it helps students to analyse the requirements for photosynthesis and how a plant is adapted for a primary production role.

We can see that a plant needs the following:

1. Carbon dioxide
2. Water
3. Light
4. Chlorophyll

From our knowledge of enzyme systems, and their control over most metabolic processes, we can also include a fifth component:

5. Temperature

From the simple equation, and some application of biological principles, we have listed five major limiting factors in photosynthesis. We will discuss limiting factors later in the book. Our concern here is how plants are adapted to get as much of these factors as possible.

2.1 The Leaf As a Photosynthetic Organ

Below is a simple diagram showing the main features of a dicotyledonous leaf. This diagram is sufficient to describe the main evolutionary adaptations of this remarkable organ:

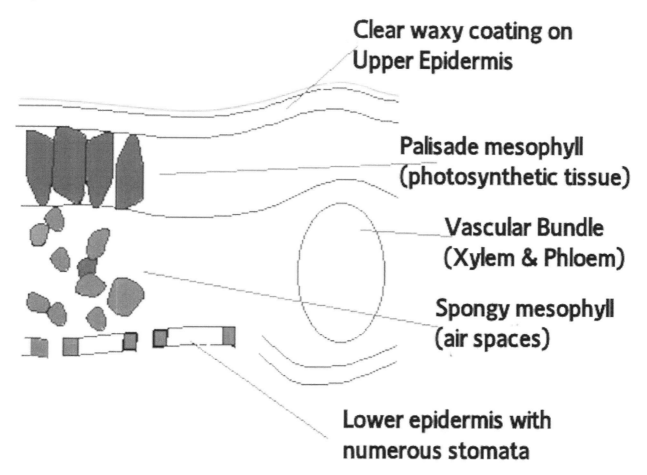

Main features:

- Clear waxy upper epidermis allows maximum light through. The waxy coating reduces water loss from the upper surface of the leaf and helps to protect it.

- Palisade mesophyll layer on upper surface of leaf. These cells are packed with chloroplasts and form the main photosynthetic tissue. Their proximity to the upper surface means maximum light reaches these important cells.

- A leaf has a branching network of veins running throughout it. These veins are the vascular bundle containing the xylem and phloem. The xylem carries water (one of the requirements of photosynthesis) to the photosynthesizing cells. The phloem in the vascular bundle carries away glucose to other parts of the plant that need it (e.g. non-green parts such as roots).

- Spongy mesophyll is loosely packed forming air spaces. This allows carbon dioxide (and other gases) to diffuse through the leaf.

- Stomata on lower epidermis allow the plant to gain carbon dioxide from the environment, and expel excess oxygen being produced in photosynthesis.

- The leaf is very thin – gases only have to diffuse a short distance to their target cells. Similarly most cells will be near to the water-supplying xylem vessels.

- The leaf has a large surface area, which means it is able to capture as much light as possible.

- Overall, these adaptations make the leaf a powerful organ for glucose production.

2.1.1. Palisade Cells

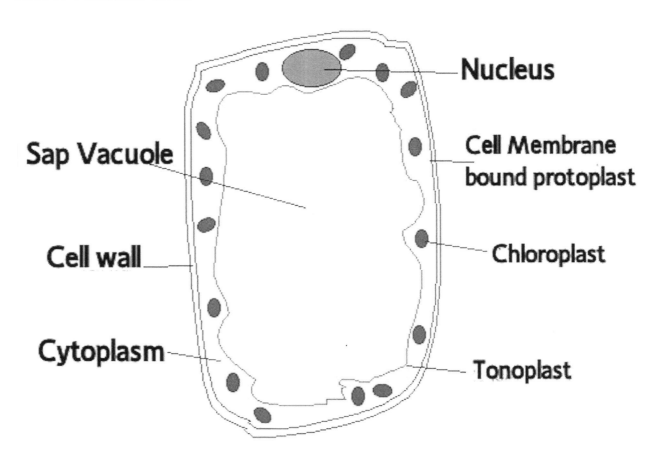

Main features:

- These are modified parenchyma cells with numerous chloroplasts.

- The cells are elongated and packed together near the upper epidermis forming a continuous layer of photosynthesising tissue.

2.1.2. Chloroplasts

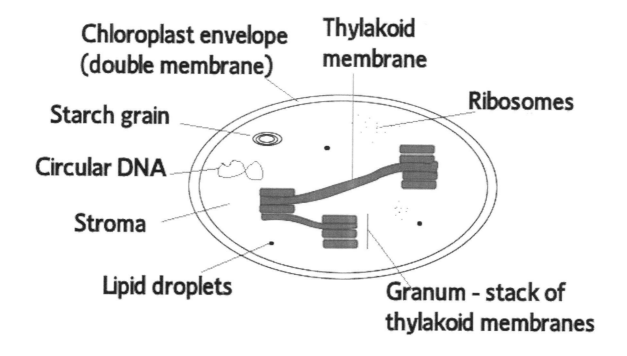

Main features:

- Vary from 1 (*Chlorella*) to about 100 in palisade cells.

- 3 – 10 µm diameter.

- Surrounded by chloroplast envelope (double membrane) – permeable to glucose, oxygen, carbon dioxide and some ions, e.g. Mg^{2+} and Fe^{2+}.

- Internal **thylakoid membrane** system contains chlorophyll and other photosynthetic pigments, enzymes and electron carriers. These flattened, fluid filled sacs, are arranged in stacks called **grana** (singular granum) – these grana are connected by **lamellae**.

- These membranes run through **stroma** which contains enzymes that catalyse the light-independent reactions.

The light-independent reactions are sometimes called the dark reactions by older videos and text books. This implies that they require the dark to proceed. This is false. In fact these reactions are independent of light, working during the day and at night. Therefore the preferred terminology is the light-independent reactions.

2.1.3. The Site of Light-dependent and Light-Independent Reactions

Students appreciate that plants are green because they capture sunlight using green chlorophyll. It is therefore easy to remember where the light-dependent, and light-independent, reactions take place:

- Chlorophyll bound to the thylakoid membranes which capture sunlight for use in the light-dependent reactions.

- Stroma does not have chlorophyll, as is does not need light. Here is the site of the light-independent reactions.

> **Thylakoid membranes carry out light-dependent reactions**
>
> **Stroma carries out light-independent reactions**

2.2 Pigments Involved in Photosynthesis

There are more pigments involved in photosynthesis than just chlorophyll. In fact chlorophyll itself is actually a group of pigments called the chlorophylls. These different pigments can be seen with a relatively straightforward chromatography experiment that students usually carry out lower down the school.

The photosynthetic pigments contain a long hydrocarbon tail, which, due to its hydrophobic nature, fits nicely into the cell membrane between phospholipids.

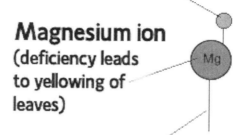

Small chemical group. This is the only difference between cholorphyll A and chlorophyll B

Magnesium ion (deficiency leads to yellowing of leaves)

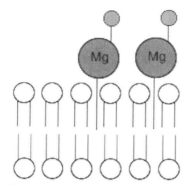

Thylakoid membrane with embedded chlorophyll molecules

Chlorophyll hydrocarbon tail

Sunlight contains all wavelengths of light from ultra violet to infrared (IR). Most of the pigments absorb only certain wavelengths, e.g. chlorophyll A and chlorophyll B absorb light in the red and blue/violet range.

Accessory pigments like carotenoids (carotenes and xanthophylls) assist chlorophyll by capturing different wavelengths of light and passing on the energy to the chlorophylls.

2.2.1. Absorption v Action Spectrums of Photosynthetic Pigments

The **Absorption spectrum** is the amount and type of light absorbed **by a pigment**, whereas the **Action spectrum** is the effectiveness of different wavelengths of light **on photosynthetic activity**

An absorption spectrum for a pigment is obtained by measuring the absorption of different wavelengths of light by the pigment. Results are plotted as a graph of absorption against wavelength.

Absorption Spectrum

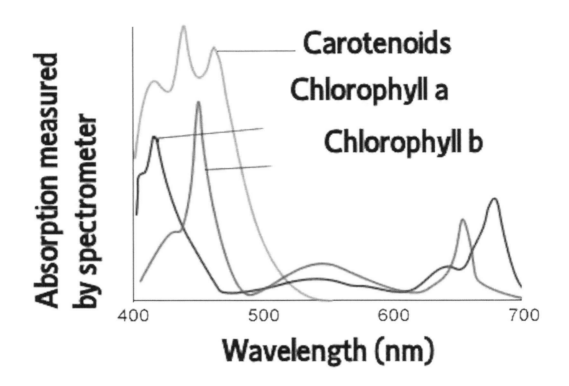

The absorption spectrum shows a lot of absorption in the blue (left), and red (right), ends of the spectrum.

Compare this to the action spectrum for photosynthesis.

An action spectrum is produced by measuring the rate of photosynthesis of a plant as it receives light of each different wavelength for a given time. Results are plotted as rate of photosynthesis against wavelength.

Action Spectrum

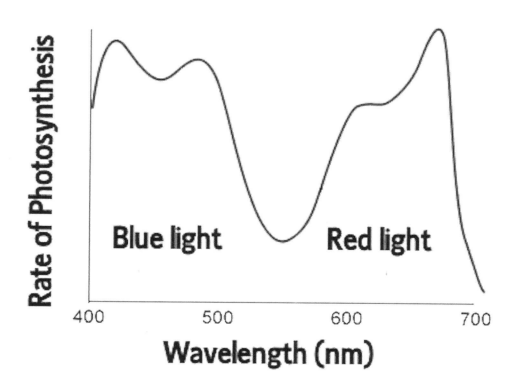

We can see that photosynthetic rate is greatest in the blue and red regions of the spectrum.

By superimposing the action spectrum with the absorption spectrum it can be seen that:

1. The wavelengths of light that are most absorbed are also the exact same wavelengths that produce the most efficient photosynthesis.

2. Different pigments are specialised to absorb different wavelengths, thereby maximizing the range of light absorbed.

With the use of simple laboratory apparatus, we can demonstrate that only some wavelengths are absorbed by a solution of chlorophyll, while others pass straight through.

Practical Idea

Grind up some dark leaves (spinach works well, though different coloured leaves could provide interesting results and be the basis for a project) in a little solvent (alcohol will do, though if you want to try chromatography on your chlorophyll sample, then 1 part 90% acetone : 9 parts petroleum ether, is a better choice). Filter to produce a crude extract of chlorophyll.

If you have a hand spectroscope (hand viewer that splits light in component colours), look through it at a bright window. The full spectrum is present. Place your chlorophyll extract in front of the spectroscope. You will see that reds and blues disappear.

If you do not have a hand spectroscope, you can use a prism to split light into the rainbow, and then place your sample in front of the prism.

IMPORTANT TAKE-AWAY: The similarity between the action spectrum of photosynthesis, and the absorption spectrum of chlorophyll, tells us that the chlorophylls are the most important pigments in photosynthesis.

3. The Two Main Reactions in Photosynthesis

There are two main groups of reactions that take place in photosynthesis - the light-dependent stage and the light-independent stage. It's important that your students don't think of them as totally separate. The light-dependent stage occurs when light is available, but the light-independent stage occurs with or without light. This means that during daylight hours, both processes are occurring at the same time. It's also important to understand that the two stages are linked. As we will see, the products of the light-dependent reactions are needed for the light-independent reactions.

3.1 Light-Dependent Reactions

Energy is transferred from light into ATP and reduced NADP (NADPH2 which is also written as NADPH + H+ in many text books, so check which syntax your own examining board uses). The reactions of the light-dependent stage can be summarized as follows:

Summary of substances entering and leaving the light-dependent reactions

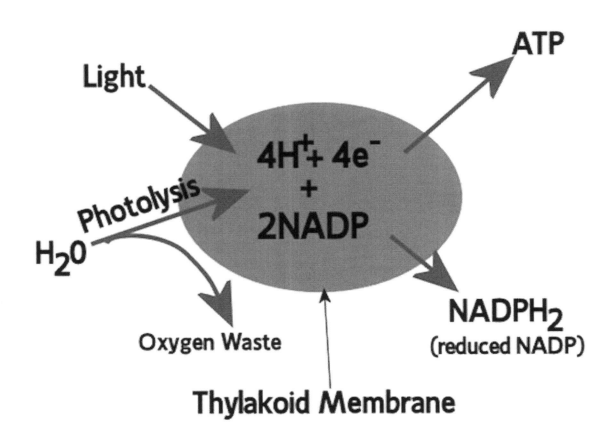

Substances entering and leaving the light-dependent stage	
Entering	**Leaving**
Light	ATP
Water	NADPH$_2$ (reduced NADP)
	Oxygen (waste)

1. Reactions of the light-dependent stage take place in the thylakoid membranes.

2. Chlorophyll and accessory pigments are arranged into two distinct photosystems – Photosystem I (PSI), and photosystem II (PSII).

3. These photosystems are visible as particles on the thylakoid membrane.

4. Chlorophyll molecules and accessory pigments in PSI absorb an average wavelength of 700nm, whereas those in PSII absorb 680nm wavelength. Therefore PSI is often referred to as P700, and PSII as P680.

The diagram below summarizes the events of the light-dependent stage of photosynthesis.

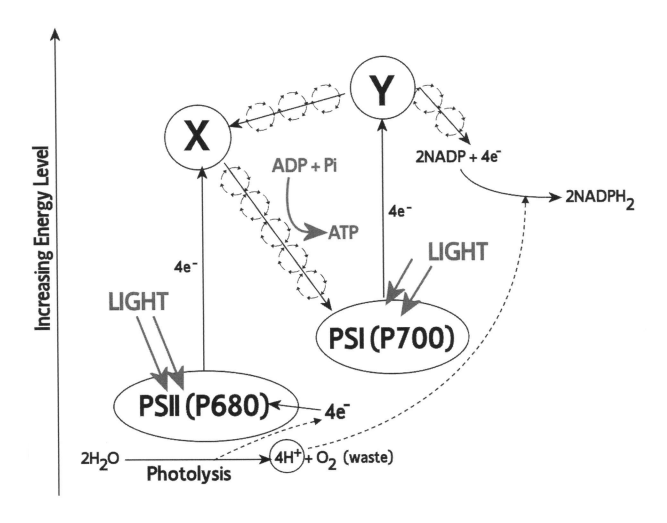

When light hits PSI and PSII, electrons are knocked out of the chlorophyll molecules. These electrons are boosted to a higher energy level. Here is the reaction for that:

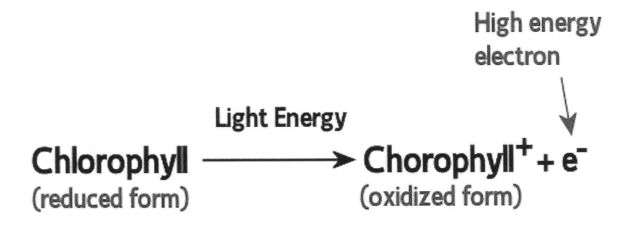

These high energy electrons do not fall back to the photosystem, but are captured by electron carriers (X & Y in the diagram above).

Result of this action:

- Electron acceptors (X & Y) are reduced.

- Positively charged chlorophyll molecule remains in photosystem.

- From X & Y, electrons are passed from one electron carrier to another (at a lower energy level). As energy is released from these redox reactions, it can be used to make ATP or reduced NADP. The pathways through which the electrons are passed can be by CYCLIC or NON-CYCLIC.

- **Cyclic returns the electrons to the photosystem from which it originated.** These electrons reduce the oxidized chlorophyll to return it to its reduced form.
  Electrons from Y are recycled back to P700 via electron carriers. During each transfer, small amounts of energy are released – this is coupled to ATP production.

- The **non-cyclic pathway means electrons do not go back to the photosystem they originated in.**

- When excited electrons leave P680 (PSII) and P700 (PSI), and are accepted by X & Y respectively, PSII and PSI are left positively charged and require a replacement electron.

- Electrons captured by X (from PSII) are passed along electron carriers to PSI, where they replace the lost electrons. Energy released as electrons are passed along is used to make ATP.

- Electrons captured by Y (from PSI) pass along carriers and combine with hydrogen ions (from photolysis of water) to form reduced NADP (NADPH$_2$).

$$2NADP + 4e^- + 4H^+ \longrightarrow 2NADPH_2$$

Here is the equation for the photolysis of water (note the electrons produced):

$$2H_2O \longrightarrow O_2 + 4H^+ + 4e^-$$

1. Hydrogen ions combine with NADP
2. Oxygen is released as waste
3. Electrons enter PSII (P680)

- Electrons lost by excitation of P680 (PSII) are replaced by those released during the photolysis of water.

- H^+ formed during photolysis combine with NADP to form $NADPH_2$ using energy released from electrons as they are passed along electron carriers from Y.

Overall efficiency of the light-dependent stage is about 40%.

3.1.1 Summary of Cyclic v Non-cyclic

	Non.Cyclic	Cyclic
First electron donor	Water	PSI
Last electron donor	NADP	PSI
Products	ATP, reduced NADPH$_2$ & oxygen	ATP only
Photosystems involved	PSI & PSII	PSI

The Effect of temperature on photosynthetic rate:

Temperature has almost no effect on the photochemical reactions involved in the light-dependent stage. However, the reactions of the light-independent stage are largely controlled by enzymes and therefore temperature does act as a limiting factor in this stage. As you expect, increasing temperature will increase the reactions of the light-independent stage.

3.2 Light-Independent Reactions

Melvin Calvin won the Nobel Prize for Chemistry in his studies on this stage. Two important points are:

1. The light-independent stage takes place in the stroma. Since the stroma contains no chlorophyll, students can easily remember that this is where the light-independent reactions occur.

2. The light-independent stage is NOT a dark reaction. It takes place in the dark and in the light. It is independent of light, hence its name.

The reactions of the light-independent stage can be summarized as follows:

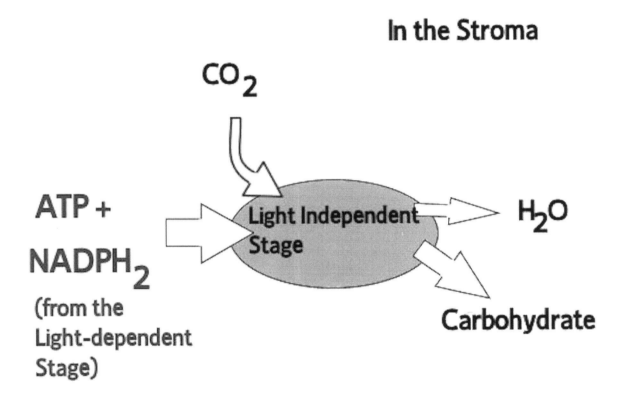

The light-independent stage gets the ATP and reduced NADP from the light-dependent stage.

> # The light-independent stage uses ATP and the reducing power of NADPH2 to reduce CO2.

3.2.1 Calvin's Lollipop Experiment

Calvin used algae called *Chlorella* in his famous lollipop experiment. He radioactively labelled carbon dioxide to use in his experiment. His idea was to trace the route taken by the radioactively labelled carbon, to see which compounds it ended up in during photosynthesis.

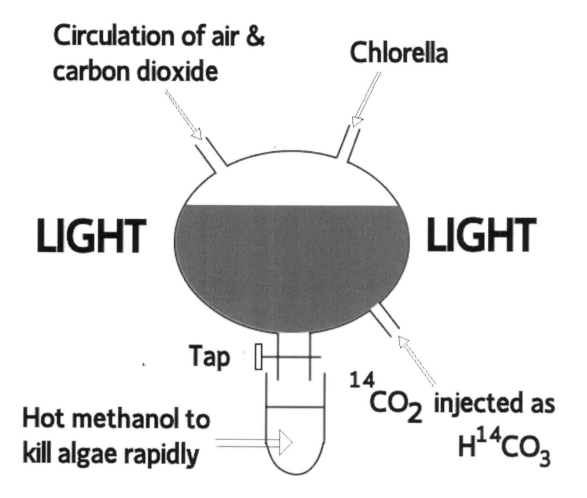

Here was his experimental approach:

1. Introduce radioactively labelled carbon-14.

2. Stop the reaction after a period of time.

3. Kill algae in hot methanol to stop reactions.

4. Extract radioactively labelled products of photosynthesis and identify using 2D chromatography and auto radiography.

- After one minute, Calvin found carbon-14 in sugars, amino acids, etc.

- After only five seconds, Calvin found carbon-14 in a 3-carbon molecule called **Glycerate Phosphate** (GP).

Using this apparatus, Calvin *et al* discovered a series of compounds that carbon-14 was built into. The route taken by carbon-14 as it is converted from one compound to another has been called the Calvin Cycle.

3.2.2. The Calvin Cycle

1. The cycle begins with carbon dioxide fixation. In the diagram below, I highlight the number of carbon atoms in each compound throughout the reaction so that you can see where they come from and where they end up.

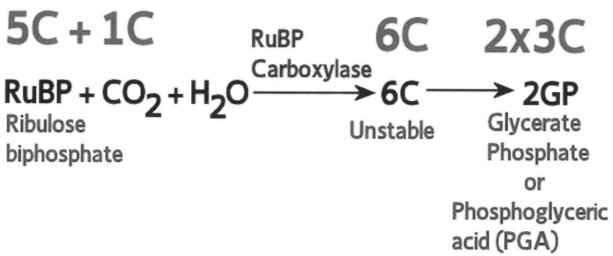

$$5C + 1C \qquad \text{RuBP} \qquad 6C \qquad 2 \times 3C$$

$$\underset{\substack{\text{Ribulose} \\ \text{biphosphate}}}{\text{RuBP}} + CO_2 + H_2O \xrightarrow{\text{Carboxylase}} \underset{\text{Unstable}}{6C} \longrightarrow \underset{\substack{\text{Glycerate} \\ \text{Phosphate} \\ \text{or} \\ \text{Phosphoglyceric} \\ \text{acid (PGA)}}}{2GP}$$

NOTE:

(GP & PGA are the same thing, so use whichever your examing board uses.
We will use GP going forward)

Carboxylation is the addition of carbon dioxide to a compound. It is carried out by a carboxylase enzyme.

2. Next, ATP and the reducing power of NADPH2 remove oxygen from GP (reducing it).

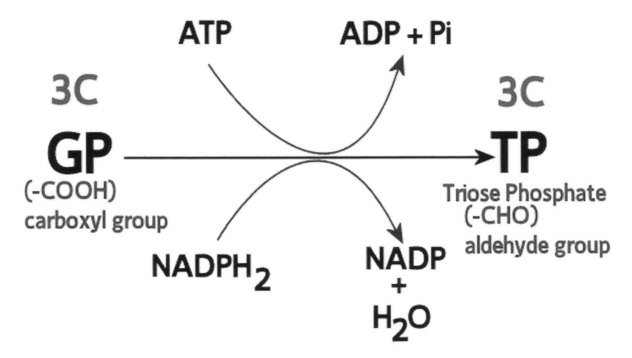

TP contains more chemical energy than GP

TP = Glyceraldehyde-3-phosphate

TP is the first stable carbohydrate formed in photosynthesis.

3. The last step in Calvin cycle is the regeneration of the carbon dioxide acceptor RuBP. **Some of the TP made in the cycle is used to regenerate the RuBP.** Some ATP is also needed.

The sequence of events leading to the regeneration of RuBP can be visualized as follows:

Light-Independent Summary
ATP and NADPH$_2$ come from light-dependent stage

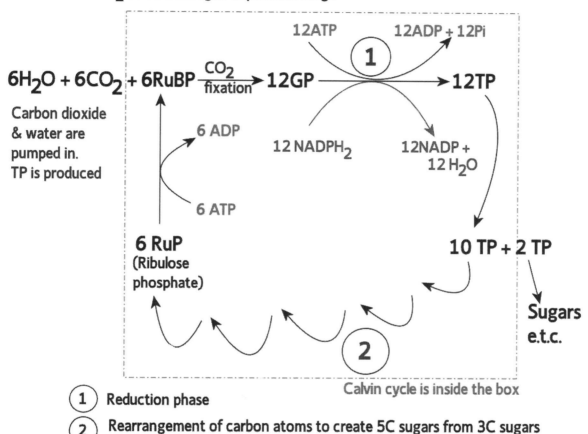

$$6H_2O + 6CO_2 + 6RuBP \xrightarrow[\text{fixation}]{CO_2} 12GP \longrightarrow 12TP$$

Carbon dioxide & water are pumped in. TP is produced

① Reduction phase

② Rearrangement of carbon atoms to create 5C sugars from 3C sugars

The CO$_2$ fixation enzyme is called Rubisco (which stands for ribulose, 1,5 biphosphate carboxylase).

Overall equation for light-independent stage is:

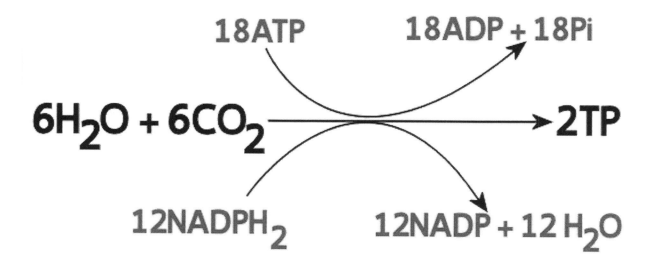

So:

6 molecules of CO₂ produce 2 molecules of TP

Simply dividing both sides by 6 to get a simpler equation of the process:

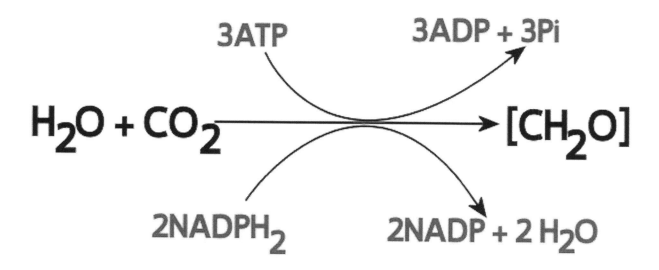

Energy in the final sugar is the same energy that originally hit the plant as light. This energy has merely been transferred from light to chemical potential energy.

3.2.3. Fate of TP

TP is immediately converted into other products so doesn't build up.

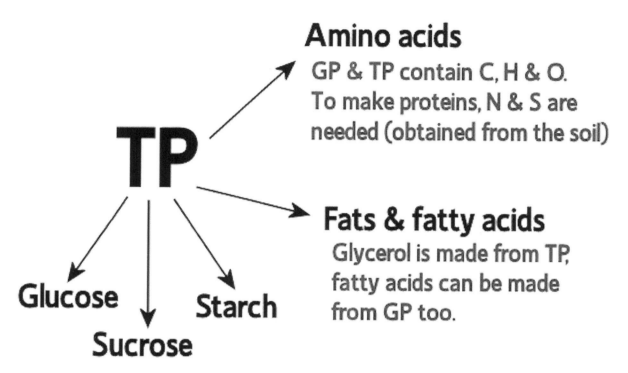

GP can be converted via Kreb's cycle to amino acids:

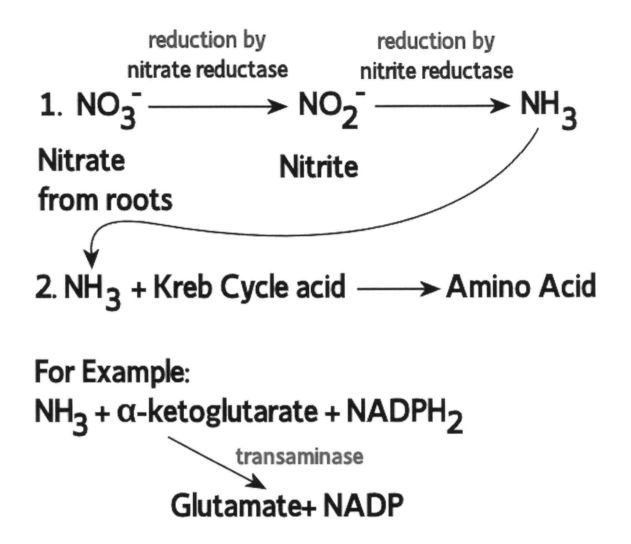

1. NO_3^- $\xrightarrow{\text{reduction by nitrate reductase}}$ NO_2^- $\xrightarrow{\text{reduction by nitrite reductase}}$ NH_3

Nitrate from roots

Nitrite

2. NH_3 + Kreb Cycle acid $\longrightarrow$ Amino Acid

For Example:

$$NH_3 + \alpha\text{-ketoglutarate} + NADPH_2 \xrightarrow{\text{transaminase}} Glutamate + NADP$$

Transamination involves transferring the amine group ($-NH_2$) from one acid to another.

3.3 Photosynthesis Summary: Linking the Light Reaction to the Tight-independent Reaction.

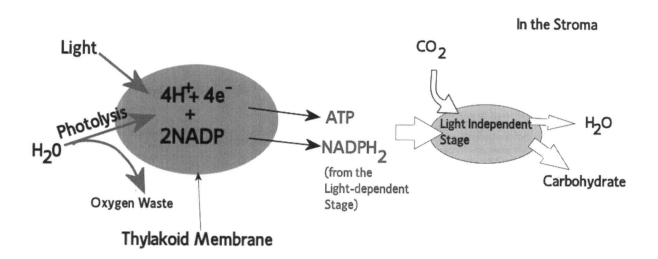

4. Limiting Factors

The principle of limiting factors:

"When a chemical process is affected by more than one factor, its rate is limited by the factor which is nearest its minimum value: it is that factor which directly affects a process if its quantity is changed."

In other words, a limiting factor is one that slows down a chemical reaction if in limited supply.

Student Question: From the equation of photosynthesis, what are the main limiting factors?

Answer: Using the simple equation:

$$6CO_2 + 6H_2O \xrightarrow[\text{chlorophyll}]{\text{Sunlight}} C_6H_{12}O_6 + 6O_2$$

.....anything to the left of the arrow head! Plus of course temperature – we now know that temperature affects the light-independent stage because it uses enzymes.

So, the main limiting factors are:

4. Light intensity

5. Carbon dioxide concentration (this is the main limiting factor on a hot sunny afternoon)

6. Water – difficult to measure though

7. Chlorophyll – some variegated leaves have cells devoid of chloroplasts – these areas appear white

8. Temperature – though this factor does not affect the photochemical light-dependent stage, though it does affect the enzymic light-independent stage

It's a good exercise to get your students to say whether these limiting factors affect the light-dependent or light-independent stages of photosynthesis.

4.1 Light

Light is obviously a limiting factor in the light-dependent stage of photosynthesis.

In the darkness, no photosynthesis can take place. Respiration does however proceed during the whole 24 hours in the day.

At night, therefore, the net gas exchange is an uptake of oxygen and release of carbon dioxide as respiration proceeds in the absence of photosynthesis (see why plants in your bedroom are a bad idea?). During this time, carbohydrates stored in the plant as starch are used up to produce energy as well as being used for growth & repair.

If a plant is left in the dark for a prolonged period (48 hours is usually enough), it becomes de-starched. This procedure is very important before carrying out experiments investigating photosynthesis using accumulation of starch as an indicator of photosynthesis.

During sunrise, light intensity increases. As it does so, the rate of photosynthesis increases. Now, both photosynthesis and respiration are occurring.

Respiration:

- Produces CO_2
- Uses O_2

Photosynthesis:

- Uses CO_2
- Produces O_2

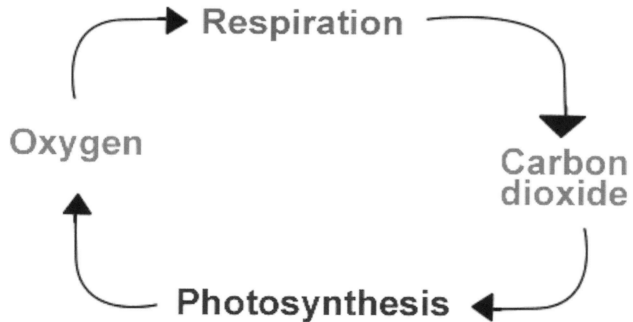

As the rate of photosynthesis increases, there comes a point when the amount of CO_2 that is being used up by photosynthesis is the same as the amount of CO_2 that is being produced in respiration. This is what's known as the compensation point.

During a normal sunny afternoon, more CO_2 is used up than produced in this manner. This point can be seen experimentally – see practical 5.4

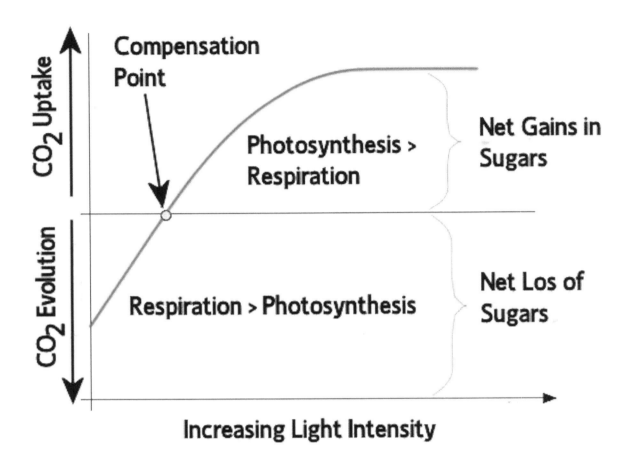

At the plateau of the curve, either light saturation has been reached, or another factor has become limiting, e.g. CO_2.

Different plants have different light saturation points. These are determined by their environment, e.g. sun and shade plants.

4.2 Carbon Dioxide

Carbon dioxide is needed in the light-independent stage of photosynthesis.

Atmospheric levels — 0.035%

Under normal conditions, this low level of CO_2 is the main limiting factor.

In agriculture, CO_2 cylinders are used in green houses to artificially increase CO_2 levels, thereby removing one of the obstacles in producing bumper crops.

 C4 plants (e.g. sugar cane) are not discussed in these teaching notes. They are called C4 plants because the first product in photosynthesis is a 4-carbon compound (as opposed to 3-carbon in the C3 plants we have discussed). C4 plants are more efficient at removing CO_2 from the atmosphere. Scientists have genetically engineered rice plants to contain C4 plant genes. These rice plants make more rice, and remove more CO_2 from the atmosphere than traditional rice plants. Is this perhaps a single solution to global warming and food shortage?

4.3 Temperature

Important only in the enzyme driven light-independent stage of photosynthesis.

Optimum for temperate plants is about 25C.

Under ideal conditions, a 10C rise doubles the rate of the reaction as $Q_{10} = 2$.

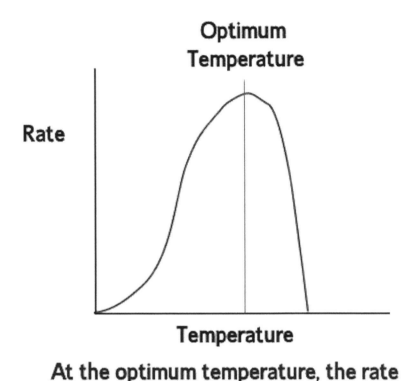

Optimum Temperature

Rate

Temperature

At the optimum temperature, the rate is at a maximum

This curve is typical for an enzyme controlled reaction. As temperature increases, kinetic energy of particles increases leading to more frequent collisions of reactants. The rate increases. There is a point where any further increase in temperature actually reduces the rate. This is because proteins, e.g. enzymes, start to denature and lose shape. They cannot do their job effectively and the rate falls off to zero.

4.4 Chlorophyll

Chlorophyll is not normally a limiting factor - plants that have insufficient chlorophyll would have been removed by natural selection. However, disease or deficiency can be a cause. A deficiency in iron, magnesium, and nitrogen can all lead to chlorosis (yellowing of leaves as a result of reduced chlorophyll), as all of these elements are required during chlorophyll synthesis.

In these situations, chlorophyll can be a limiting factor in the light-dependent stage.

Lack of light can also cause chlorosis, as light itself is needed for chlorophyll synthesis. This explains why grass goes yellow when garden furniture is not moved frequently enough.

The herbicide dichlorophenyl dimethyl urea (DCMU), stops non-cyclic electron flow in chloroplasts; therefore inhibiting the light-dependent reactions. What effect would it have on $NADPH_2$ and ATP?

4.5 Water

It is not possible to measure the direct effect on photosynthesis of water, as all cells require water for many processes.

5. Practical Work

To measure the effects of limiting factors, apparatus can be bought from the usual suppliers (like Phillip Harris). However, a cheaper option is to get your students to make their own. They can then be responsible for maintaining and looking after their own piece of apparatus during their practical work.

The diagram below shows the final assembly of a simple photosynthometer:

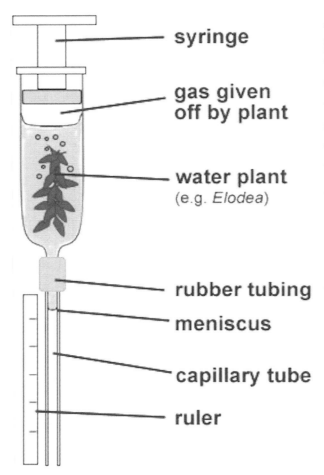

syringe

gas given off by plant

water plant
(e.g. *Elodea*)

rubber tubing

meniscus

capillary tube

ruler

As gas is given off, it collects under the piston of the syringe. This in turn increases the pressure inside the syringe which forces water down the capillary tube. The faster the rate of photosynthesis means the faster the water movement along the capillary tube.

Students can measure the rate of movement of the meniscus.

In all experiments outlined below, students should add a pinch on hydrogen carbonate powder to the water. This ensures that CO_2 is unlikely to be a limiting factor (water plants can use hydrogen carbonate as a substrate for photosynthesis).

5.1 Measuring the Effect of Light on the Rate of Photosynthesis

The basic idea of this experiment is to vary the distance of the light source from the apparatus, and measure the corresponding rate of photosynthesis.

Things to note:

1. A light source can heat up the water in the apparatus. We know that heat can speed up reactions through increased kinetic energy of particles, more collisions and so forth. Therefore there needs to be a heat shield. Students can immerse the apparatus in a beaker of water as this works well.

2. Experiments should be carried out in a darkened room, so that the only light the plant gets is the light provided by the experiment.

3. Let the plant acclimatize to the new conditions before starting the measurement of rate.

4. For graphs, students need to know that light intensity is inversely proportional to the square of the distance. $I \propto 1/d2$. This means that if distance is doubled, light intensity is 1/4. For each distance, work out light intensity as $1000/d2$, where d is the distance in cm between plant and light source.

5. Results can be plotted as light intensity against rate.

5.1.1. Extending This Practical

1. If you are feeling adventurous, you can calculate the percentages of oxygen and carbon dioxide in the gas bubble given off by the plant. See practical 5.5 for details of this technique.

2. Discuss the effect respiration had on the results.

3. Consider looking at compensation point of the plant. See practical 5.4 for details.

4. The apparatus used is a little crude. My students always enjoy researching and coming up with alternatives. They then list the benefits and drawbacks of each piece of apparatus.

5.2. Carbon Dioxide

The basic idea is to make up different concentration of sodium hydrogen carbonate, and then use these solutions instead of water in the apparatus. I would suggest making up a range from 0.01 mol dm^{-3} to 0.10 mol dm^{-3}.

Things to note:

1. Plot a graph of concentration of hydrogen carbonate against rate.

2. While CO_2 concentration is being varied, all other factors must be kept constant. It is worth going over these with students and discussing how to achieve them – light and temperature are the most important.

3. Temperature doesn't just affect the rate. Air expands as it is heated which could cause problems in measuring rate if students handle the apparatus during measurements.

5.3 Temperature

Measuring the effect of temperature should not pose too many problems. You can use the same apparatus and water baths (or beakers maintained at constant temperatures). Rates of photosynthesis can then be measured and plotted against rate.

Things to note:

1. Students need to maintain constant levels of CO_2 and light at each temperature. A suitable light intensity and CO_2 level can be determined from earlier experiments.

2. At each temperature, the apparatus needs to be set up and left for a while to equilibrate. Remember that gas expands as it is heated so an erroneous reading for rate will be obtained if this is not done.

5.4 Compensation Point

Photosynthesis results in a net uptake of CO_2 and production of O_2.

Respiration results in a net uptake of O_2 and production of CO_2.

In low light, respiration is working faster than photosynthesis. As you gradually increase light intensity, there comes a point when respiration exactly balances photosynthesis in the exchange of these gases. This is the compensation point. Increase light further, and photosynthesis works faster than respiration with a net output of oxygen.

Now, measuring compensation requires a different apparatus to the one used so far, but if anything, it is simpler:

Leaf from terrestrial plant

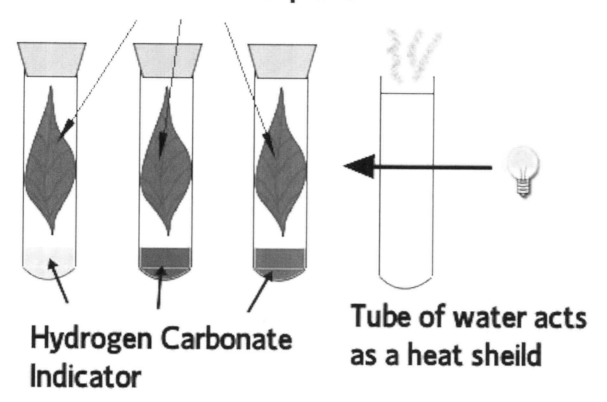

Hydrogen Carbonate Indicator

Tube of water acts as a heat sheild

Things to note:

1. Setup the apparatus as shown in the diagram with several tubes, each a different distance from the light source.

2. Each tube needs a heat shield to absorb heat from the light so that the experimental tubes do not warm up.

3. Leave the apparatus set up for at least two hours, and preferably over-night (gently shake periodically if possible) before analysing results.

5.4.1 Preparing Hydrogen Carbonate Indicator Stock Solution:

Dissolve 0.2g thymol blue and 0.1g cresol red in 20cm³ ethanol. This is the dye solution. Then dissolve 0.84g sodium hydrogenbcarbonate in 900cm³ of distilled water. Add the dye solution to this solution and make up the volume to 1 litre.

To Use:

1 part stock solution: 9 parts distilled water.

Colour changes:

Carbon dioxide given off during respiration increases acidity.

Yellow   Orange   Red   Purple

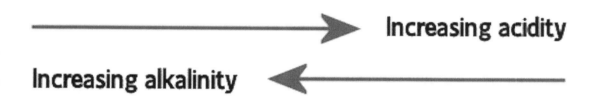

Increasing acidity

Increasing alkalinity

At atmospheric levels of CO2, hydrogencarbonate indicator is cherry red.

Prior to use, hydrogen carbonate indicator should have atmospheric air bubbled through it until it reaches the cherry-red colour.

Analysing results:

CO_2 given off will dissolve in the indicator making indicator more acidic.

CO_2 removed from indicator makes indicator less acidic or more alkaline.

1. If conditions become more acidic, this can be interpreted as respiration rate being greater than photosynthesis.

2. If conditions become more alkaline, this can be interpreted as photosynthesis rate being greater than respiration.

5.5 Gas analysis

To measure the percentages of carbon dioxide and oxygen in a gas sample given off by a plant (or exhaled air in mammals), you first need a sample of gas. Set up a photosynthesis experiment using the ideal conditions found in previous practical work so that a suitable amount of gas can be collected. The gas, when pushed through a capillary tube, must be several centimetres in length.

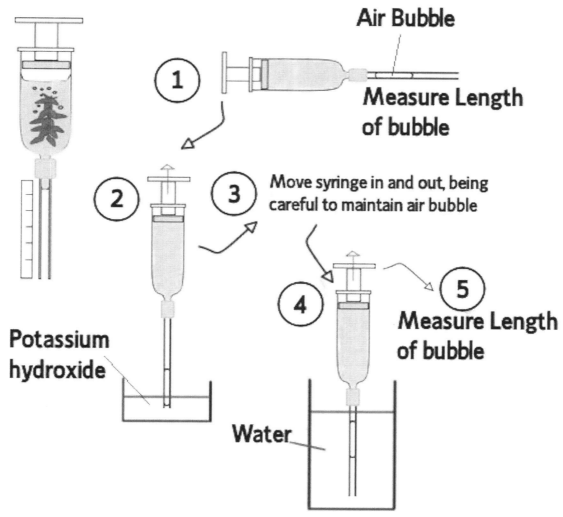

Now repeat 1 to 5 using pyrogallol instead of potassium hydroxide in step 2.

1. Following your photosynthesis experiment, invert your syringe and push some of the air into the capillary tube. Ensure it is flanked by water on either side. Put the capillary tube immersed in a beaker of water at room temperature for

five minutes. This allows the air to reach a particular volume at that temperature. Measure and record the length of the bubble. It should be several centimetres long. Keep the beaker of water for stage 4.

2. Push the bubble to within 1cm of the end of the capillary and then draw up potassium hydroxide solution until the bubble is within 1cm of the syringe end.

3. Still holding the capillary tube in the end of the potassium hydroxide, move the bubble back and forward in the capillary tube several times slowly as you go. This brings the air into contact with potassium hydroxide clinging to the wall of the capillary tube. Potassium hydroxide absorbs carbon dioxide.

4. Put the capillary back into water for five minutes. This gives time for carbon dioxide to be absorbed.

5. After five minutes, measure the length of the bubble. Record the new length. Remember, any decrease in length is due to removal of carbon dioxide.

Now, repeat steps 2 to 5 using pyrogallol instead of potassium hydroxide. Pyrogallol reacts with potassium hydroxide in the tube to produce potassium pyrogallate. Potassium pyrogallate absorbs oxygen. Record the final length of the capillary tube bubble after oxygen removal.

> # Warning: pyrogallol is harmful and needs to be handled with care.

These readings will give you:

1. Length of carbon dioxide bubble, and
2. Length of oxygen bubble

Using symbols:

a = Original length of bubble

b = Length after CO_2 removal

c = Length after O_2 removal

Therefore:

$$\text{Percentage } CO_2 = \frac{a - b}{a} \times 100$$

$$\text{Percentage } O_2 = \frac{b - c}{a} \times 100$$

5.5.1. Where to Go Next?

If students get accurate measurements with this technique, it is the good basis of a project to look at gas composition given off by aquatic plants in photosynthesis under different light intensities. What are the results going to be at the light compensation point?

6. Downloads from my website

I've created an examination you can use in your classroom to fully test your students on the topic of photosynthesis. The exam lasts approximately 50-60 minutes, and I've included a full mark scheme.

Because an exam is not much use to you on a Kindle device, I've setup a download where you can get your copy. I've created a PDF version for easy printing, but I have also included it in Microsoft Word format so you can add/edit the test yourself if you want to.

DOWNLOAD EXAM & IMAGES HERE

http://goo.gl/fzUG3

If you wish to make the exam longer, I suggest perhaps an essay style question. Essays test students on their ability to formulate ideas, and put them into coherent sentences in a logical and informative way. Not all exam boards use essays in their examinations, but even so, I feel they are a valuable tool for teachers and students alike.

More Advanced Biology Teaching Notes?

If you liked this book and found it helpful in your teaching of Photosynthesis, please leave a review on Amazon. This will help me decide whether or not to continue this series of Teaching Notes. In your comment, please leave ideas of what you would like to see next – 'Respiration' perhaps?

Review on Amazon here:

http://www.amazon.com/dp/B00BMV0LV4

https://www.amazon.co.uk/dp/B00BMV0LV4

In other Amazon stores, search for **B00BMV0LV4**

Name: _____

Date: _____

Percentage:
Grade:

Photosynthesis Exam

1 hour

Answer all questions.

Write answers on the exam paper.

Scores:	
1	
2	
3	
4	
5	
6	
7	
Student TOTAL:	
Exam Total:	

Question 1:

Briefly describe the main adaptations of a leaf which make it the ideal photosynthetic organ. **[6]**

_____

_____

_____

_____

_____

_____

_____

_____

_____

_____

_____

_____

_____

_____

Question 2:

(a) State the site of:

    (i)   The light-dependent reactions (photophosphorylation) **[1]**

_____

    (ii)   The light-independent reactions (Calvin cycle) **[1]**

_____

(b) In the space below, draw a fully labelled chloroplast **[4]**

(c) (i) It has been suggested that chloroplasts have evolved from free-living photosynthetic organisms. Suggest *two* pieces of evidence from the structure of a chloroplast to support this theory. [2]

_____

_____

(ii) What other cell organelle has similar characteristics that could mean it also was once free living? [1]

_____

Question 3:

(a) (i) What is an absorption spectrum? [2]

_____

_____

_____

_____

(ii) What is an action spectrum? [2]

_____

_____

_____

_____

(iii) How are the two linked? [2]

_____

_____

(iv) Why do plants have several different photosynthetic pigments? **[2]**

Question 4:

The diagram below shows the photosystems involved in the light-dependent stage of photosynthesis.

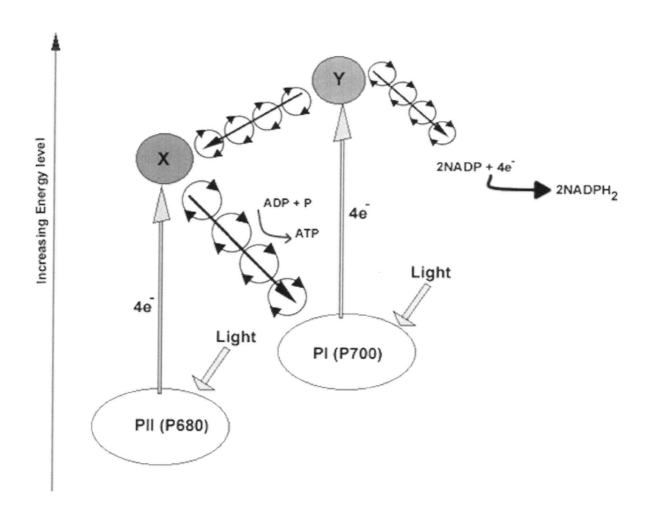

(a) (i) During the reactions, water is split. What is this splitting called?[1]

_____

(ii) Write the equation for the splitting of water. [3]

_____

(iii)   Complete the diagram by drawing on what happens to the products of the splitting of water. **[4]**

(b) (i) What is meant by cyclic photophosphorylation? **[2]**

_____

_____

_____

_____

_____

_____

(ii) What is meant by non-cyclic photophosphorylation? **[3]**

_____

_____

_____

_____

_____

_____

_____

_____

Question 5:

(a) What links the light-dependent and the light-independent reactions? **[2]**

_____

_____

_____

(b) Below shows an incomplete summary of the light-independent process:

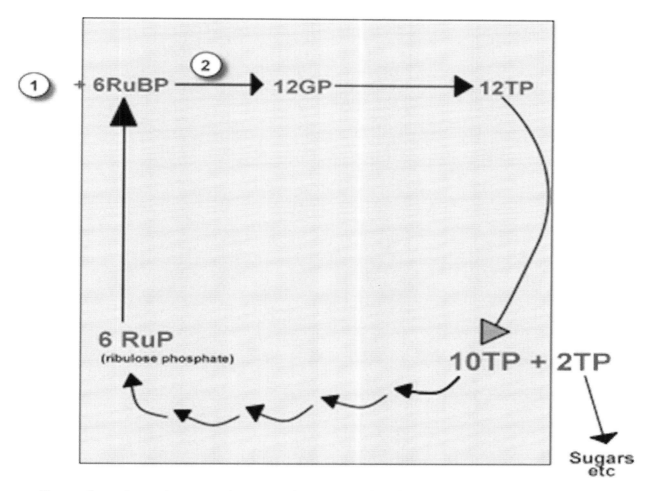

(i)   Complete the equation at point (1) on the diagram **[1]**

(ii)   What is the process at point (2) called? **[1]**

(iii)   What is the significance of the shaded box? **[1]**

_____

(iv)   Draw on the diagram where ATP is used, and where the reducing power of NADPH₂ is used. **[3]**

Question 6:

(a) (i) What is a limiting factor? **[2]**

_____

_____

_____

_____

_____

_____

(ii) How could you tell if a factor in one of your experiments was limiting? **[2]**

_____

_____

_____

_____

_____

_____

(iii) State 5 possible limiting factors in photosynthesis? **[2]**

_____

_____

_____

(b)(i) What is the light compensation point? **[2]**

_____

_____

_____

_____

_____

_____

(ii) How would you expect the light compensation point to differ in plants that grow in bright light and those that grow in the shade? Explain your answer. **[3]**

_____

_____

_____

_____

_____

_____

_____

_____

Mark Scheme

Question 1: (*1 mark for each of the points, MAXIMUM 6 marks*)

1. Clear waxy upper epidermis allows maximum light through.

2. Palisade mesophyll layer on upper surface of leaf. These cells are packed with chloroplasts and form the main photosynthetic tissue. Their proximity to the upper surface means maximum light reaching these important cells.

3. Vascular bundle contains xylem to bring the photosynthesizing cells water – one of the requirements of photosynthesis.

4. Phloem in the vascular bundle carries away glucose to other parts of the plant that need it (e.g. non-green parts such as roots).

5. Spongy mesophyll is loosely packed forming air spaces which allow carbon dioxide to diffuse through the stomata (holes in the lower epidermis).

6. Stomata on lower epidermis allow the plant to gain carbon dioxide from the environment and expel excess oxygen being produced in photosynthesis.

7. The leaf is very thin – gases only have to diffuse a short distance to their target cells. Similarly most cells will be near to water-supplying xylem vessels.

8. Leaf has a large surface area meaning that it captures as much light as possible.

TOTAL 6 Marks

Question 2:

(a) (i) Thylakoid membranes (Not Chloroplasts) [1]

(ii) Stroma of chloroplast [1]

(b) (*½ mark for each of the points, MAXIMUM 4 marks*):
1. Diagram that fills space available;
2. No "hairy" lines;
3. Scale indicating around 3-10 µm in diameter;

The following drawn and labelled:
1. Chloroplast envelope;
2. Thylakoid membranes
3. Ribosomes
4. Granum

5. Lipid droplet
6. Stroma
7. Circular DNA
8. Starch grain

(c) (i) (*1 mark for each of the points, MAXIMUM 2 marks*):
1. Similar size to prokaryotes;
2. Have their own DNA, suggesting self-replication;
3. DNA is circular like bacteria;
4. 70S ribosomes are the same sort typical of prokaryotes (Eukaryotes has 80S);

Note: Chloroplasts reproduce by binary fission as do bacteria. This is not a valid point in the answer though as it is not directly related to structure of chloroplasts.

(ii) Mitochondria [1]

TOTAL 9 Marks

Question 3:

(i) Amount of light absorbed [1]; at different wavelengths of light [1];

(ii) The rate of photosynthesis [1]; at different wavelengths of light [1];

(iii) The 2 spectra are similar, indicating that the pigments that absorb light [1]; Are responsible for the photosynthesis we observe [1];

(iv) Different pigments absorb different wavelengths of light [1]; making absorption more efficient [1]; and allowing photosynthesis in different lighting conditions [1];
MAXIMUM 2

TOTAL 8 Marks

Question 4:

(a)(i) Photolysis [1]

(ii) Suggestion for marking – award 3 marks for the equation, then take off one mark for every error.

$$2H_2O \longrightarrow 4H^+ + 4e^- + O_2$$

(iii) H^+ : Completing the equation and showing an arrow from photolysis to position of this equation:

$$2NADP + 4e^- + 4H^+ \rightarrow 2NADPH_2 \quad \text{(2 Marks)}$$

e^-: Showing electrons entering photosystem II (1 Mark)

O_2 : Written, or implied as waste material (1 mark)

(b)(i) (*1 mark for each of the points, MAXIMUM 2 marks*):

Cyclic photophosphorylation occurs in PSI only [1];

Electrons from PSI return to PSI;

Energy released from transfer is coupled to ATP production (photophosphorylation;

(ii) (*1 mark for each of the points, MAXIMUM 3 marks*):

Electrons do not return to photosystem they originated from.

Electrons from PSII pass to PSI via electron carriers;

These electrons replace those lost by PSI;

Energy is lost on the way, which is coupled to ATP production;

Electrons from PSI pass along carriers and combine with hydrogen ions and NADP to form reduced NADP;

TOTAL 13 Marks

Question 5:

(a) ATP and reduced NADP [1] from the light-dependent are the "fuel" for the light-independent [1].

(b) (i) & (ii)

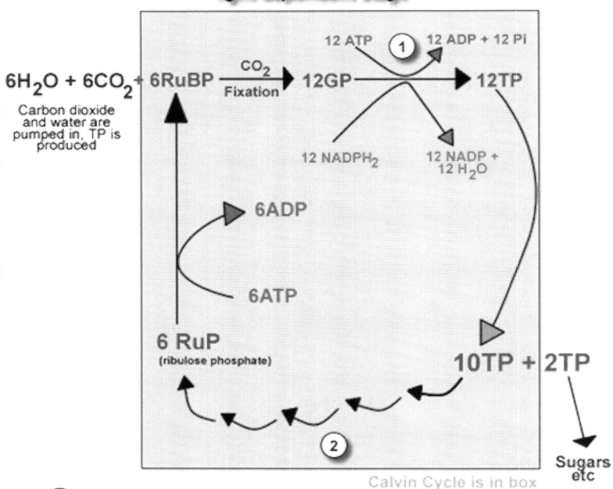

Light-Independent summary: ATP and $NADPH_2$ come from light-dependent stage

$6H_2O + 6CO_2$ + 6RuBP — $\xrightarrow[\text{Fixation}]{CO_2}$ → 12GP —→ 12TP

Carbon dioxide and water are pumped in. TP is produced

12 ATP ① 12 ADP + 12 Pi

12 $NADPH_2$ 12 NADP + 12 H_2O

6ADP

6ATP

6 RuP
(ribulose phosphate)

10TP + 2TP

②

Sugars etc

Calvin Cycle is in box

① Reduction phase

② Rearrangement of C atoms to generate 5C sugars from 3C sugar

(iii)   Represents the light-independent reaction.

(iv)   See diagram.

TOTAL 6 Marks

Question 6:

(i) A factor that reduces the rate of a reaction [1], if concentration is too small when compared to other reactants [1].

(ii) Increase the amount of the factor [1], and if the rate increases, that factor was limiting [1].

(iii) 2 marks for all 5. 1 mark off for each missing.
1. Light
2. Water
3. Temperature
4. Carbon dioxide
5. chlorophyll

(b)(i) As the rate of photosynthesis increases, there comes a point when the amount of CO_2 being used up by photosynthesis [1], is the same as the amount of CO_2 being produced in respiration [1]. This is known as the compensation point.

(ii) Plants that grow in lower light conditions have lower light compensation points [1]. That is lower light levels are required to produce the same rate of photosynthesis [1]. This is so they can compete with other plants [1] for available light. For example, a plant growing in the shade of a tree, requires less light to photosynthesize than the tree [1]. Any three points, or other valid explanations.

TOTAL 11 Marks

Exam Total Marks 53 Marks

Printed in Great Britain
by Amazon.co.uk, Ltd.,
Marston Gate.